FLORA OF TROPICAL EAST AFRICA

RESEDACEAE

J. ELFFERS AND P. TAYLOR

Annual or perennial herbs, rarely shrubs. Leaves scattered or fasciculate, simple to pinnatipartite ; stipules small, gland-like. Inflorescence terminal, spicate or racemose. Flowers hermaphrodite or rarely unisexual, usually irregular. Calyx persistent or deciduous, 4–8-partite. Petals mostly 4–7, usually unequal, small, deciduous or persistent, free or slightly coherent, laciniate, simple or clawed, often having a membranous appendage at the base of the limb. Disc usually present, often dilated on the adaxial side. Stamens 3–40, perigynous or inserted on the disc, often declinate, free or monadelphous at the base, not covered by the petals in bud ; anthers 2-celled, introrse. Ovary superior, sessile or stipitate, of 2–6 free or connate carpels. Ovules 1–∞, inserted on parietal placentas or at the base of the ovary. Fruit a closed or open capsule, indehiscent, rarely baccate or of as many follicles as carpels. Seeds mostly numerous, reniform or hippocrepiform, exalbuminous, with a curved embryo.

Carpels free almost to the base, each with 2–3 basal ovules ;
fruit composed of 5–6 follicles, connate at the base only . 1. **Caylusea**
Carpels joined to form a unilocular ovary with parietal
placentas ; fruit a capsule . . . . . . 2. **Reseda**

1. CAYLUSEA

St.-Hil., 2e Mém. Réséd. 29 (1837), *nom. conserv.*

Annual or short-lived perennial herbs, often branched, prostrate or erect, glabrous or hairy. Leaves entire, linear to elliptic, flat or undulate. Flowers densely racemose, white ; pedicels bracteate at the base. Calyx 5-partite. Petals 5, deciduous, inserted at the base of the disc, simple or laciniate, dilated inside at the base of the limb into a membranous appendage. Disc elevated, patelliform, fleshy, adaxially dilated. Stamens 10–15, inserted on the disc ; filaments papillose or hispid ; anthers declinate, 2-celled, dehiscing longitudinally. Ovary stipitate on the disc, of 5–6 whorled carpels, connate at the base ; ovules numerous, clustered at the carpel-bases. Follicles open, spreading stellately, ciliate-hispid on the margins. Seeds few.

A small genus of three species, the third widespread in North Africa and extending into the Sudan and Eritrea.

Stems scabrid ; leaves linear to lanceolate ; adaxial petals
4- or 5-fid ; seeds rugulose . . . . . 1. *C. abyssinica*
Stems pubescent ; leaves lanceolate to elliptic ; adaxial
petals 2- or 3-fid ; seeds densely and minutely tuberculate . . . . 2. *C. latifolia*

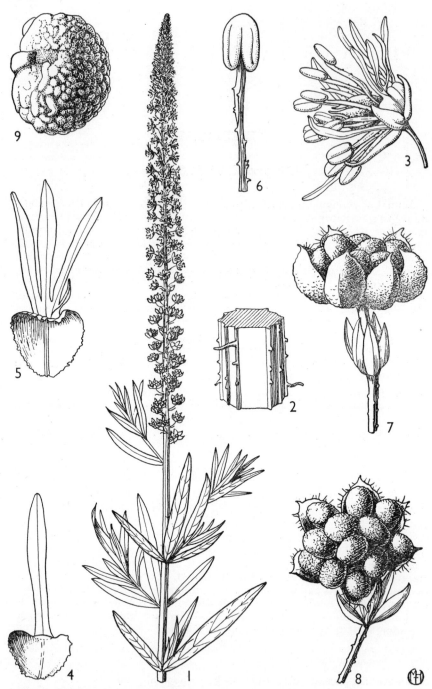

FIG. 1. *CAYLUSEA ABYSSINICA*, from *Milne-Redhead & Taylor* 10076—**1,** part of plant, × 1 ; **2,** part of stem, × 8 ; **3,** flower, × 8 ; **4,** abaxial petal, × 16 ; **5,** adaxial petal, × 16 ; **6,** stamen, × 16 ; **7 & 8,** lateral and dorsal views of fruit, × 8 ; **9,** seed, × 24.

1. **C. abyssinica** (*Fresen.*) *Fisch. & Mey.* in Ind. Sem. Hort. Petrop. 7 : 43 (1840) ; Muell. Arg. in Mon. Réséd. : 229 (1857) ; F.T.A. 1 : 103 (1868); J. Elffers in K.B. 1955 : 630 (1956); P. Tayl. in K.B. 1958 : 286 (1958). Type : Ethiopia, Adowa, *Schimper* 103 (K, neo. !).

Erect or ascending herb, occasionally bushy, 1·5–10 dm. high. Stem simple or branched, striate, glabrous or often scabrous to sparsely hairy on the angles, especially towards the inflorescence. Leaves linear or lanceolate, 2–8 cm. × 2–18 mm., scabrid on the nerves and margins, occasionally undulate. Racemes terminal, dense, spike-like, 5–40 cm. long. Bracts linear-subulate, persistent. Pedicels slightly shorter than the bracts but elongating considerably in fruit. Sepals subequal, up to 2·5 mm. long, persistent. Petals white, up to 3·5 mm. long, 2 adaxial deeply 4- or 5-fid, 3 abaxial entire or 2-fid. Filaments papillose ; anthers salmon-coloured or orange, turning yellow at maturity. Fruit about 4 mm. in diameter, 7–12-seeded. Seeds hippocrepiform, about 1·4 mm. long, rugulose with small rounded warts which are fused to form transverse wrinkles, pale brown when ripe. Fig. 1.

UGANDA. Acholi District : Chua, Agoro, 12 Nov. 1945, *A. S. Thomas* 4355 ! ; Karamoja District : Mt. Morongole, June 1942, *Dale* U246 ! ; Kigezi District : Kabale, Mar. 1950, *Purseglove* 3348 !
KENYA. Northern Frontier Province : Moyale, 9 Oct. 1952, *Gillett* 14020 ! ; Uasin Gishu District : Soy Road, Aug. 1931, *Brodhurst-Hill* 196 ! ; Nairobi District : Thika Road House, July 1951, *Verdcourt* 552 !
TANGANYIKA. Moshi District : Lyamungu, S. slope, Kilimanjaro, 24 Aug. 1932, *Greenway* 3139 ! ; Dodoma District : Kazikazi, 13 May 1932, *B. D. Burtt* 3615 ! ; Mbeya, 1 Mar. 1932, *R. M. Davies* 349 !
DISTR. U1, 2 ; K1, 3–6 ; T2, 5–7 ; Eritrea and Ethiopia
HAB. A weed of secondary grassland, abandoned cultivations, waste places, riversides, etc. ; 1200–3000 m.

SYN. [*Reseda pedunculata* R. Br. in Salt, Abyss. App. 64 (1814), *nomen nudum*] *R. abyssinica* Fresen. in Mus. Senck. 2 : 106 (1837)

2. **C. latifolia** P. *Tayl.* in K.B. 1958 : 285 (1958). Type : Kenya, Huri Hills, *Mrs. J. Adamson* 628 (K, holo. !, EA, iso.)

Erect herb or subshrub, branched from near the base ; stems striate, glabrous at the base, densely pubescent above. Leaves lanceolate to elliptic, acute or acuminate, up to 6 × 2 cm., entirely glabrous or sparsely papillose on the midribs. Racemes terminal, very dense, spike-like, up to 20 cm. long. Bracts narrowly lanceolate, acuminate, persistent, about 1·5 mm. long. Pedicels slightly longer than the bracts, elongating in fruit. Sepals subequal, up to 2 mm. long, persistent. Petals white, about 4 mm. long, 2 adaxial deeply 2- or 3-fid, 3 abaxial entire. Filaments papillose ; anthers yellow. Fruit about 5 mm. in diameter, 7–8-seeded. Seeds hippocrepiform. about 1·2 mm. long, densely and minutely tuberculate, black when ripe.

KENYA. Northern Frontier Province : Huri Hills, July 1957, *Mrs. J. Adamson* 628.
DISTR. K1 ; not known elsewhere
HAB. Volcanic soil, probably in semi-desert scrub ; 1500 m.

2. **RESEDA**

L., Sp. Pl. : 448 (1753) & Gen. Pl., ed. 5 : 207 (1754)

Stephaninia Chiov., Fl. Somala : 77 (1929)

Erect or decumbent herbs or subshrubs, glabrous or pilose, more or less branched. Leaves simple, entire or variously dissected. Inflorescence a terminal, spike-like raceme ; flowers very numerous. Calyx 4–7-partite, Petals 4–7, free, unequal, longer than the torus, simple or clawed, 2–∞-fid.

Fig. 2. *RESEDA ELLENBECKII*, from *Gillett* 13300—**1**, part of plant, × ½ ; **2**, leaf, × 1 ; **3**, young flower with a sepal, and some petals and stamens removed, × 15 ; **4**, abaxial petal, × 24 ; **5**, adaxial petal, × 24 ; **6**, fruit, × 3.

often dilated on the inside to form a membranous appendage. Disc sub-sessile, cylindric or urceolate, fleshy, often dilated adaxially. Stamens 10–40, inserted on the disc. Ovary sessile or stipitate, unilocular, open and 2–4-lobed at the apex (very exceptionally closed), obtusely angled ; placentas 3–6, parietal, multi-ovulate ; ovules sessile, pendulous, numerous, arranged in 2–4 rows. Capsule polymorphous, open and 2–4-lobed at the apex, dry, membranous or coriaceous, many-seeded, with a few exceptions indehiscent. Seeds rotund or reniform.

This large genus is concentrated mainly in the Near East and in countries surrounding the Mediterranean.

Reseda odorata L., the common Mignonette, which is widely cultivated for its delicate fragrance, is native to North Africa and is often grown in gardens in East Africa where it may occasionally occur as a garden-escape.

Leaves entire, glabrous, ovate, lanceolate or obovate, 5–12 cm. long, 1·5–5·5 cm. broad ; fruit a capsule, closed and obscurely 2-toothed at the apex   .   .   1. *R. ellenbeckii*
Leaves divided into filiform or narrowly linear seg-ments, glabrous or ± scabrid, 2–4 cm. long ; seg-ments acute, channelled, 0·5–1·5 mm. broad ; fruit a capsule, open and 3-toothed at the apex   .   .   2. *R. oligomeroïdes*

1. **R. ellenbeckii** *Perk.* in E.J. 43 : 417 (1909). Types : Ethiopia, Audo Mt., *Ruspoli & Riva* 994 (FI, isosyn. !) ; "Gallahochland," Tarro Gumbi, *Ellenbeck* 2093 (B, syn.†)

Erect herb or subshrub, simple or branched. Leaves alternate, petiolate with petioles 0·5–2 cm. long ; blades ovate, oblong, lanceolate or obovate, 5–12 cm. long × 1·5–5·5 cm. broad, narrowed into the petiole, shortly acumi-nate or acute, rigidly membranous, entirely glabrous. Raceme elongate, densely spike-like, 10–35 cm. long ; bracts 3–6 mm. long, linear, caducous. Flowers 1·5–4 mm. long ; pedicels ± 1 mm. long, papillose, lengthening in fruit. Sepals 5, equal, 2 mm. long, lanceolate, acute, soon deciduous. Petals 5, unequal, 1–2 mm. long, broadly obovate, 1- to 5-fid, soon deciduous. Stamens 10–11, 2–3 mm. long ; anthers basifixed ; filaments deciduous. Ovary obovate and compressed below, drawn out into a club-shaped or cylindrical tube above, very shortly 2-toothed at the apex. Capsule pendu-lous, obovate, up to 1·7 cm. long × 1 cm. broad, compressed or slightly inflated, very shortly mucronate and closed at the mouth. Seeds 8–11, reniform, about 1 mm. long, densely tuberculate, black when ripe. Fig. 2.

KENYA. Northern Frontier Province : Daua River, Murri, 27 June 1951, *Kirrika* 98 ! and Lag Hareri, 23 May 1952, *Gillett* 13300 !
DISTR. **K**1 ; also Ethiopia
HAB. Alluvial ground or under trees near watercourses ; 400–660 m.

NOTE. This is an anomalous species, being very distinct in the nature of its fruit. It is not inconceivable that it may ultimately be placed outside the genus *Reseda*, but for the time being it appears satisfactory to include it in the broadest sense of this genus.

2. **R. oligomeroïdes** *Schinz* in Bull. Herb. Boiss. 3 : 397 (1895). Type : Somaliland Protectorate, "Tuju steppe," *Keller* 52 (K, iso. !)

Erect branched shrub ; old stems with thin, pale brown, shallowly fissured bark and prominent leaf-scars. Leaves very numerous and densely crowded, 2–4 cm. long, glabrous or ± scabrid, forking into 3–5 filiform to narrowly linear segments ; segments 0·5–1·5 mm. wide, channelled on the upper surface and acutely pointed. Raceme densely spike-like, 3–20 cm. long ; bracts filiform, about 4 mm. long, caducous ; pedicels about 1 mm. long, elongating to about 2 mm. long in fruit. Sepals 6, equal, about 2 mm.

long, linear to narrowly oblong, deciduous. Petals 6, very unequal, 2–3 mm. long, broadly obovate, 1- to 5-fid, deciduous. Stamens about 14, 2–3 mm. long. Ovary narrowly obovoid, open and deeply 3-toothed at the apex. Capsule erect, obovoid, ± 3-angled, about 7 mm. long. Seeds 20–50, reniform, 0·5–1 mm. long, smooth, black when ripe.

KENYA. Northern Frontier Province : S. end of Lake Rudolf, 20 Aug. 1944, *Mrs. J. Adamson* 28 !
DISTR. **K**1 ; also Somaliland Protectorate and Somalia.
HAB. Volcanic soil, probably in semi-desert scrub ; 360 m.

INDEX TO RESEDACEAE